MORE
PRIZEWINNING
SCIENCE FAIR
PROJECTS

MORE PRIZEWINNING SCIENCE FAIR PROJECTS

PENNY RAIFE DURANT

SCHOLASTIC INC.

New York Toronto London Auckland Sydney

ISBN 0-590-02936-3

Copyright © 1998 by Penny Raife Durant.
All rights reserved. Published by Scholastic Inc.
Scholastic and associated logos are trademarks and/or registered trademarks of Scholastic, Inc.

12 11 10 9 8 7 6 5 4 3 2 1   8 9/9 0 1 2 3/0

Printed in the U.S.A.   40
First Scholastic printing, March 1998

For my nieces and nephews — Elizabeth, David, Monica, Greg, Becky, Alison, and Gretchen Raife — with much love

Table of Contents

Acknowledgments

Without the scientific expertise of Paul Mitschler, this book would not have been possible. I am grateful for his help and enthusiasm. I would also like to thank my agent, Robin Rue, who is too wonderful for words and more than understanding. I am indebted to Ann Reit, my editor, for being a gracious and helpful guide.

Introduction

Congratulations! You've entered into an exciting and challenging endeavor: the Science Fair! You're joining millions of others who have done science fair projects in the past as well as young Americans all over the nation doing projects now.

There are many good reasons for doing a science fair project. You will learn how to research a topic, how to organize your work, how to experiment, and to use scientific skills. You may be thinking that you'd like to pursue a career in science. You may not. Some high school students win money and scholarships every year with their science fair projects.

All the skills you will learn will help you, no matter what you decide to do as a career. Plus, it will be interesting, because you get to choose what you want to study.

Scientists help us identify problems and search for solutions to the problems. A survey is one way to gather information. Research is another way.

Experimenting is a third. In your science fair project, you may be doing these same things. If you conduct a survey and find out that your schoolmates don't know the basic facts about AIDS or lung cancer, you are identifying a problem. You may have some suggestions on how to remedy the problem. You won't be finding a cure for AIDS or cancer, but you might be saving one of your friends from getting one of these diseases by suggesting that better information be given to all the students in your school. You could save a life!

1.
Safety

The most important thing you need to think about before you do any work on your experiment is *safety*. Some chemicals and solutions are toxic and can harm you or others. Any time you use electricity, you must be careful. Here is a list of safety precautions. Read this list carefully before you begin ANY project.

1. NEVER WORK ALONE. Check with your adult partner before you take any actions by yourself. While many science fairs will not allow more than one student to work on a project, you will need an adult partner.
2. Check your equipment carefully. What are potential hazards? Are you working with glass? How can you be sure it won't break? Look for sharp metal edges as well.
3. Make sure all electrical equipment is functioning properly before you begin. Check the cords before you plug in anything. Make sure

they are not frayed or cracked. Be sure your power supply is grounded.

4. Check with your science teacher or a chemist before you work with chemicals or acids. Combinations of harmless chemicals can be volatile or toxic.

5. Don't keep extra chemicals, acids, or other solutions around. If you have small children in your house, see if you can have a locked cabinet for your science things. Check with your science teacher about proper ventilation and temperatures for stored material.

6. Use goggles to protect your eyes from chemicals and particles. Wear a rubber apron to protect your clothes. Ask your teacher if you need to wear rubber gloves.

7. Find out and follow proper disposal techniques for chemicals, agar plates, and chemical solutions. Most of them should not be put down the drain or in the trash.

8. Check with your science teacher to see if you need special permission from the region or state science fair if your project involves people, animals, chemicals, DNA, or bacteria.

9. Treat your subjects — humans or animals — humanely.

10. Check with your science teacher before you put anything in your display. Photos of your work will be safer than actual equipment.

11. You will need expert help if you plan to do a blood-related project. You will also need authorization from science fair officials.
12. Use a laboratory for chemical experiments.
13. Clean up after yourself.
14. The more complicated and dangerous your experiment, the more you will need help. Ask your adult partner to stay with you and watch.
15. If you are not sure if something is safe, check with your adult partner first.

2.
So, What Is a
Science Fair Project?

First of all, a science fair project asks a question. That question can be answered by studying or experimenting. "Why should I wash my hands?" or "Does shaving body hair help a swimmer go faster?" are questions that can be answered by studying or experimenting. They could be science fair projects. "Will my grandmother come to visit?" cannot be answered by studying or experimenting. It would not be a good science fair project.

There are four types of science fair projects: an experiment, a research project, a survey, and a model. Most science fairs do not want models, but the first three are welcome. If you want to do a model anyway, realize that your score will not be very high.

A science fair project involves several steps, no matter what you choose to study. Just as professional scientists do, you will follow the steps of the *scientific method* to experiment, to research your topic, or to conduct a survey. After you have done your work, you will create a display for judges and the public to see what you did and what you learned. Don't panic! All the information you'll need to know about the scientific method and how to create a display will be in this book. All the suggested projects ask questions. You may want to follow one that's in the book. You may want to choose something entirely different. That's fine. Either way, you'll follow the same steps.

3.
Creating a Time Line

It's important to create a time line for your work. Write down the date of your science fair. This will be the last date on your time line. Now write down today's date. How much time do you have between the two dates? If it is just a couple of days, you have waited far too long. Ideally, you should have about three months, though some projects may take longer.

You will need time to choose a topic and have your project approved by your teacher, time to find an adult to work with you, time for your research, and time to carry out your experiment. You may need extra time to conduct your experiment again, if you think something went wrong. Then you will have to analyze your results, draw conclusions, and put together a display to show your school and judges what you did.

This may sound like a lot of time needed, but if you work slowly and steadily, you will be able to get everything done without panicking. Give yourself deadlines along the time line. Make sure you plan to finish the experiment at least two weeks before the science fair. That will give you time to work with your results and make the display.

Whatever you do, don't wait until the last minute!

Here is a sample time line:

October 15: Choose topic, submit it to teacher for approval.
October 17: Begin research.
November 10: Design experiment.
November 10–December 10: Conduct experiment(s). This may have to be changed if you are conducting an experiment that will take longer.
December 10: Collect final data.
December 15–31: Work with your results, draw conclusions.
January 1: Begin work on presentation and display.
January 15: Science Fair

4.
The Scientific Method

Don't be alarmed by the term *scientific method*. This is a method that many scientists use to work in an orderly fashion. By following the steps of the scientific method, one scientist can repeat the work of the first to see if the results are the same. When one scientist makes a remarkable discovery, colleagues across the world need to repeat the work to see if they also get the same results. If enough find the same results, then the discovery is proven. If a scientist has not been careful, then the results, no matter how exciting they may be, are not automatically accepted by other scientists.

The scientific method is an outline, or plan, that helps you work in an orderly fashion. If you follow this plan, other scientists (teachers and judges) will be more impressed with your work than if you do not follow it. Use it as a guide to help you.

The Purpose: The purpose of your project should be fairly simple to state. Let's say you are surveying the students in your school to see how aware they are of the link between smoking and cancer. You might state your purpose like this: The purpose of this project is to determine how aware students in Kennedy School are of the link between smoking and cancer. The purpose tells anyone who reads it exactly what you wanted to know.

The Hypothesis: The hypothesis is your educated guess at the answer. It is a possible solution to the question that your project asks. After you've studied the effects of smoking on the human body, found out how many teens start smoking every day, and read all you can about the link between smoking and cancer, you will have to guess about your schoolmates. Have you had classes that taught about the effects of smoking? If teachers have been telling you for years about smoking's link with cancer, do you think your classmates will understand it? Will it show up in your survey? Make a guess. Then you will test that guess.

The Procedure: The procedure is the next step in the scientific method. You will have done research, as described above. You've kept a project notebook, even if you aren't conducting your experiment in a science lab. Now it's time to design your experiment. In the case of surveying your classmates about smoking, you will have to develop very specific questions about smoking and cancer. You will also have to know how many students you must ask to get a good sample. You'll take all these things into consideration when you design your survey. It would be the same if you were growing one group of soybeans under an incandescent bulb and another group under a Gro-Lite. You must design the experiment to be as accurate as possible. Write down every step you follow to get your results.

The Results: After you've conducted your experiment, or survey, or research, you will have to report what you found out. It's very important to keep track of everything in your lab notebook. You may not report everything you write down, but it will be good to have it just in case you need it or someone asks to see it.

Results are usually reported in numbers. Often you will use charts or graphs to show measurable results. If you use weights and measures, use metric measurements. Scientists use metric measurements worldwide.

The Conclusion: Your conclusions are based on your results. What percentage of students didn't know of the link between smoking and cancer? How many know of the link but think they will smoke anyway? This is the place where you accept or reject your hypothesis. It is okay if you have to reject your hypothesis. It doesn't mean you didn't do your job. Many hypotheses are rejected. Scientists take this information and begin again with a new hypothesis.

In the conclusion, you will try to explain why you accepted or rejected the hypothesis. What does it mean?

If you follow the scientific method, putting together your project will be easier. It will also make your display easier to construct. And, your results are more likely to be accurate. Let the scientific method work for you.

5.
Choosing a Topic to Study

You will be spending a lot of time with your project. Choose a topic you find very interesting. Look over the titles in this book or look through your science book at school. What sounds interesting to you?

Another way to choose a topic to study begins with identifying a problem. There are many problems in our world. Listen to a single newscast and you'll be able to list several. Some problems are vast, like the destruction of the rain forests, world hunger, or global warming. You will not be able to address such a huge topic in your science fair project. However, you might be able to take a small part of a problem and study that piece.

For example, you might see a lot of cars rushing down the freeway or thruway. Most of the cars look the same. There are vans, sedans, station wagons, eighteen-wheelers, buses, etc. But most of the sedans look alike from a distance. All of them use gasoline, a nonrenewable resource, and all of them emit carbon dioxide, which contributes to the greenhouse effect.

If you study car design, you will know that vehicles that are more aerodynamic than others get better gas mileage, use the fuel more efficiently, and therefore emit less carbon dioxide. Taking this information a step or two further, could you design a car that is aerodynamic? How would your car be different from the other cars now on the market? If you research automobile designs and build a wind tunnel to test your design, you will understand why both the auto manufacturers and the consumers of cars need to be serious about these things.

A science fair project is an information-gathering tool. Research, surveying, and experimenting are all ways to gather information. You will have to decide which method is best for obtaining information on the question you want to ask.

Where can you find other problems to study? There are a number of sources. News magazines have science sections that present current findings and studies. Science magazines present research as well. Your newspaper prints articles that reflect problems in our world. Listen to your friends and parents talk. What is a problem for them? Can you think of a scientific way to look at the problem or a part of it?

Narrowing your topic to a single, answerable question is the next step. Let's say you want to study animals. That's a huge topic. Narrow it a bit. Dogs interest you. That's better, but still too large. Let's say you have an old dog, a mutt you suspect is part collie. Your dog knows how to beg for food and let you know when he needs to go out. Have you heard the expression, "You can't teach an old dog new tricks"? Is it true? You can put together a science fair project to test this question. You would research training dogs and make a plan, following the scientific method.

Finding an adult partner is another step in deciding which science fair project to do. If you want to study nuclear fusion, you will have to find an adult partner who not only knows about nuclear science, but one with access to the equipment you would need. This simply might not be possible.

Try to pick a realistic project for someone your age, with your resources. Your uncle may be a dog breeder. He would be an excellent partner for a project about teaching old dogs new tricks. Your parents are the most likely people to be your partners. Discuss your ideas with them.

6.
Researching Your Topic

Research is vital to your project. The more you know about a subject, the more accurate your answer to your question will be. Research will help you narrow your subject. When you begin to learn about dog training, you may decide to try to train your dog to do a new trick without using food as a reward.

Step one in researching is to organize your work. Use a *project notebook*, where you record everything you do. It doesn't need to be fancy. It can be a spiral-bound notebook or a loose-leaf notebook or lined pages stapled together with a cover.

When you read a book or an article, write down in your notebook the name of the author, the title, the date published, and who published it. This way, if you have to go back to find something you missed, you will be able to find it. Use a different page for each resource. Put the publishing information at the top and write your notes below. This will help you write your bibliography.

Start your research with your school library. You will find books and magazine articles that should be helpful. Is there a list of resources the author used? You might be able to find those, too. Your public library will have information in both the children's area and the adult area. They may have a computerized system for finding articles on your subject. Ask the reference librarian for help. Another good source of information is a college or university library.

If you have access to the Internet, you will find a wealth of information on-line. You will use a *search engine*, or a program that brings information to your computer when you give it a subject for which to search.

Pick out several *key words* for your project. Keep a list of them and add others as you find them helpful. Try using them to pinpoint information. The more specific you can be, the more helpful articles you'll find. Some sites will be lists of Internet addresses you might need. You can place a *bookmark* at these and return again and again for more information.

In a recent search, I found 1,054 sites for Global Warming, 684 for Deforestation, and 3,969 for Recycling. That is far too many articles to read or print out. See if you can combine key words and focus in on your subject a bit more.

Make sure you let the search engine know you want to combine the key words. When searching for information about gray water, I forgot to do that; the search engine found 95,765 sites for gray *or* water. When I told it to include both terms, gray *and* water, it found 644. Different procedures are needed for different programs. Check with the owner or operator of the computer you are using to see how it works.

Print out articles or parts of articles that will be of help to you. At the top of each printed page, your computer should put the on-line address of your source. You can go back and find it again if you need it. If you can't print out the information, take notes in your project notebook and make sure to write down the on-line address.

If you are in middle school or junior high, you will be expected to read information from adult journals and periodicals. If you don't understand something you read, jot it down and ask your science teacher or adult partner about it. Use your notebook for asking questions, too. Note things that pop into your mind as you read. "Why are most of the gray water articles from Arizona?" is something you might wonder. Jot the question down. You may know the answer by the time you have read the articles, but you may have to seek out more information.

You may need to go to a professional for some information. If you want to train your dog, you might want to talk to a puppy trainer. You would find someone like this in the yellow pages of your phone directory. Or ask at the local pet shop. If you have done some research and can ask intelligent questions, you will find professionals very helpful.

Research will probably help you to see what you need to carry out your project. Are the materials you need going to be expensive? Does your school have materials and equipment you can use? Knowing what you may have to buy is important before you get too far into your project. Talk it over with your adult partner.

Research will be a part of your project from the beginning until the end. Keep reading and learning more about your subject, even after you've designed and are conducting your experiment. The more you know, the better your project will be.

7.
The Experiment

Are you ready to plan your experiment? You will be ready if you've done your research and you have an adult partner. It's time to refine your idea. Let's say you've decided to ask, "Is Gray Water Good for My Plants?"

Because you've done your research you know what plants need. You know that gray water is water that has been used in your shower, sink, or washing machine. Some houses are designed so this water is collected, not washed down the sewer with toilet water. For your experiment, you'll have to decide how to collect gray water without re-plumbing your home. You also know why this is an important question for areas of our country and the world where water is scarce.

VARIABLES
Everything you can change about a project is a *variable*. You will need to list all the variables as-

sociated with your project. Let's list the variables for growing plants with gray water:

Plants need water, light, soil, food. But there are variables within these needs. If we focus closely on the variables within the needs of a plant, we come up with the following list:

- amount of water
- amount, intensity, and color of light
- type and quality of soil
- temperature of environment and soil
- type and amount of food (fertilizer)
- watering pattern, method, and frequency

In addition, we must consider the type and age of the plant. There are many things we can do to experiment with plants. By changing just one variable, say the color of light a plant receives, we can design an experiment to test just that.

In order to test the effect of gray water on plants, we need to keep all the variables the same for our plants. They must receive the same amount, color, and intensity of light. They must have the same soil, the same amount and pattern of watering, the same amount (if any) of fertilizer, and the plants must be the same type and size. The only thing we will vary is the type of water given.

USING A CONTROL

In order to show that the gray water makes a difference, we need a control. This means we will use a plant that has everything just the same as the test plant. We will water one with gray water and one with regular tap water. The plant receiving ordinary tap water is the control. It shows what would have happened without the experimental factor of gray water.

MEASURING

When you plan your project, you need to think about what you will measure. If you're planning to survey your friends to see how many of them know someone with cancer, you will work with numbers. Even with plants you need to work with numbers. You need to measure something. These numbers you get will be your data. You will work with the data when you present your results. Measure everything you add to the plant. Write down in your notebook: "Wednesday, November 2. Gave 7 ml of gray water to plant *A* and 7 ml of tap water to plant *B*."

Do you want to weigh the root systems? If so, you will cut off the root systems after the experiment is completed and rinse the roots. Weigh just the root systems. You might prefer to count the number of new leaves or measure the size of new leaves. In your notebook and report, you will look at the health of the plants, noting any brown leaves, spots, the color of the new leaves, etc., but you need to measure something in an objective way.

HYPOTHESIS

After you have done your research, you should have a good idea about your experiment. You're ready to make a hypothesis. A hypothesis is an educated guess at what will happen in your experiment. A possible hypothesis for the gray water experiment is: Plants treated with gray water will put out as many new leaves as plants treated with tap water.

This hypothesis tells what you think will happen. It also tells that you are going to count the number of new leaves. Of course, if you are going to count the number of new leaves, you will need to know the number of old leaves. It's best to take photographs of your work. We'll talk about this more when we talk about displays. But it's good to think of it now, before the experiment starts.

Now you're ready to set up your experiment. You know what your plants need. You know what you will do differently with each plant or group of plants. Remember, only one variable! You know what you are going to measure. All of this is very specific. Make sure you write it down in your project notebook. Go over this with your adult partner or science teacher.

MATERIALS

What do you need to carry out your experiment? If you go over this chapter, you will know what you need for the gray water experiment. Write down everything you think you will need. Check with your adult partner. Is anything on your list too hard to get? Too expensive? Too complicated to build? This is a good time to change your experiment if you have to do so.

PROCEDURE

Write down each step of your plan. List them in order in your notebook. The first step will be buying the plants or seeds you want to use. Note the kind and where you bought them. Write down the type of soil you are using, the type of containers, and where you are going to put the plants. How often are you going to water them? How much water at a time will you give them? How will you make sure they get sufficient light and heat?

When you make your display, you will write up your procedure in simple steps, even though you have much more information in your notebook. You will simplify things for the display. Keep detailed notes in your notebook.

Plan to give yourself enough time for the experiment to show results. Plants can take a long time to grow. Talk with a nursery person about what type of plant might show results sooner for the type of experiment you want to perform. Don't try to conduct a plant experiment in a week. You won't be giving the experiment a fair trial.

RESULTS

You've given your experiment enough time. You've made a record of everything you have done so far. Now you can collect the data and show your results. At this step, you will measure what you said you would measure. If you decided to weigh the root systems, you will cut them from both plants and carefully rinse away the soil. Then, using a scale, you will weigh each and note it in your notebook. If you decided to count the new leaves, you will do just that.

This is an important step in your experiment. Accuracy of data collection is vital. Use the right tool to measure what you want to measure. Be careful when you take measurements and when you write them in your notebook.

CONCLUSION AND INTERPRETATION

Now is the time to answer your question. What happened? You researched your topic. You know a lot about your subject. What do the results mean? If the measurements you made show a difference of four new leaves, what does this mean? How healthy are the gray water plants? What would you suggest for others to try in an experiment like yours? Is there anything you did that might have influenced your results? This would include anything that went wrong, like the dog leaving open the door to the outside where your plants were located and frost nipping the leaves. It might include anything you would do differently the next time.

The most important thing you will do in this step is to decide if your data support your hypothesis. Did the gray water plants develop as many new leaves as the tap water plants? Is gray water good for plants?

8.
Conducting a Survey

You may choose to do a project that involves conducting a survey instead of an experiment. You will still follow the scientific method. As with any project, you will begin with research. Decide what you want to know. Maybe you want to know if students in your grade level eat breakfast on school mornings. Your research will show you the importance of eating breakfast. It might even show you national percentages of children who do and do not eat breakfast.

HYPOTHESIS

After you have done some research, you will put together an educated guess as to what percentage of the students in your grade level eat breakfast on school mornings. You may think your school is average and use the same numbers as the national average. You may think that *everybody* eats breakfast. Whatever you think, state it something like this: Ninety-nine percent of the students in my grade level eat breakfast on school mornings.

SETTING UP A SURVEY

Before you decide to conduct a survey at your school, you will have to ask permission. Talk to your science teacher and see if you need written permission from the principal.

After you get permission, decide what your population will be. In the example we're using, the *population* is the students in your grade level at your school. The people who answer your survey are your *sample*. The larger your sample, the closer to a true picture you will get. If you give a written sample to all the students in the sixth grade, you will not get all of them back. Some people do not want to answer survey questions. Some students will forget. Some may lose them.

In order to get honest answers, it is best not to know who wrote the answers. Let the students answer anonymously. Some answers you get may be silly. You may have to throw them out.

Next you will need to work on your questions. You need to make sure that what you ask is very clear. Ask only what you want to know. If you want to know if the students eat breakfast, ask it in a clear manner. "Do you eat breakfast before coming to school?" would be one way to ask, if your school doesn't have a breakfast program. But if students might eat breakfast at school before classes begin, you will not get good results. "Do you eat breakfast on school days?" might be a better way to word the question. Do you want a yes or no answer? What about the students who usually eat breakfast, but not always? Do you want to ask if they always eat breakfast or give an option of answering some- times as well as yes or no? Work on these ques- tions with your adult partner. Try them out on several people who are not in your survey popula- tion just to see if they understand them.

Then you will either ask the questions or hand out papers with the survey questions on them. If you hand out papers, be sure you let the popula- tion know when you need them back. It might be best not to let them take the surveys home, just to eliminate lost or forgotten surveys.

RESULTS

You will count answers for your results. Figure out what percentage of the sample eats breakfast. For your display, you will want to make a graph. There are several types of graphs. Circle graphs and bar graphs are two types. A computer can help you make the graphs after you know your percent- ages.

CONCLUSION AND DISCUSSION

Results are numbers. You use them to accept or reject your hypothesis in your conclusion. Then you discuss why you think you got the numbers you did. Think about anything that might have gone wrong while you were conducting your survey. If nothing appeared to go wrong, say so. You can accept the results.

Some surveys are given in person or over the telephone. You may want to ask your neighbors a survey question. Whatever you decide to do, talk it over with your parents or adult partner first. Some neighborhoods are too busy or unsafe for you to go out in alone. You may have to adjust your survey to fit your situation.

9.
The Presentation

There are three parts to your presentation: the report, the display, and the judging.

THE REPORT
Scientists build on the work of other scientists. If they didn't, they would have to repeat everything anybody has ever done in their field. Once certain things are proved, scientists can build on them. They learn about other scientists' work through reports.

Scientists structure their reports in certain ways to make them clear and easy for other scientists to follow. That is what your report should do. Another student should be able to take your report and do the same experiment.

The report begins with an Abstract. If you are in middle school or junior high, your teacher and the judges may expect you to have an Abstract. It goes on top of your report, but it is usually written last. An Abstract is a brief account of what you did and what you found out. Don't take more than 250 words, about a page typed double-spaced, to write the Abstract. It must include your hypothesis, a short description of your experiment, and your conclusion.

The Title Page comes next. Most science fairs do not want the name of the student to appear on the projects or reports. You will be assigned a number or space. Leave your name off your report.

The title is the first thing people will see on your display. Make it interesting. Many people like to use a question for a title. All the titles in this book are questions. Or, you may want to word your title something like, "The Effect of Gray Water on Houseplants." Check with your science teacher to see if you should include scientific names in your title.

After the Title Page comes a Table of Contents. This is something you will write later, after you see how many pages your report will take. Judges find this a help. They may want to read your conclusions in the report. Your Table of Contents page will help them find it quickly.

A page of Acknowledgments comes next. Here you thank people who helped you and give them credit for their help. Your adult partner, professionals who gave you information, maybe even the librarian or your teacher who helped you would be listed here. Simply thank the people by their full names and say what they did. "I want to thank Dr.

Bill Sanchez, for his help in designing the experiment and in providing the scales I used to weigh my results."

The *experiment report* tells everything you did. You will organize it much like the experiment was organized. It includes the Purpose, Materials, Procedure, Results, Conclusion and Discussion, and a Bibliography.

The Purpose is a statement that tells in a very simple way what you intended to do with your work. "The purpose of this experiment was to see if I could teach my old dog a new trick," is an example.

The Materials you used must be listed in your report. It is important to be accurate, especially if you used scientific equipment from a laboratory. Scientists must be careful to provide drawings of equipment they have developed and used in their experiments. That way others can replicate their experiments and help prove their results.

For the Procedure, you will list each of the steps you followed in your experiment. Show how you controlled all the variables except the one you were testing. Mention the control if you used one.

The Results come next. Remember that results deal with measurements or numbers. Results are data. They don't prove or disprove your hypothesis. You will need to work with the numbers to see if they are significant. Check with your science or math teacher to see if you should run a test on your numbers to see if the differences are significant.

In the Conclusion and Discussion you answer the original question you asked. Can an old dog learn new tricks? You might have taught your dog to shake hands. In this part of your report, you would talk about what happened and how it caused you to accept or reject your hypothesis. For example, you might write, "I didn't think my ten-year-old Labrador retriever could learn a new trick, but he learned to shake hands in just one week. The old saying, 'You can't teach an old dog new tricks' isn't true."

If your dog didn't learn a new trick, you might mention factors that influenced this to happen. "My ten-year-old Labrador retriever is only interested in sleeping and eating. Every time I tried to teach him something, he just repeated his one trick, lying down."

Or if something interrupted your experiment or caused a problem, this is the place to mention it. When he was in kindergarten, my son wanted to see if plants could grow in cat food. We made a mixture of cat food and soil and planted seeds. However, before the seeds had a chance to germinate, our dog ate the experiment! We had put the plants outside to get some sunlight. They were in a box on the picnic table. The dog jumped onto the table and ate the cat food, leaving only the control in potting soil. Disaster can strike any experiment. If it is too late to redo your experiment, use the conclusion and discussion part of your report to explain what happened.

This is also a place to tell what you would do differently the next time. If you didn't control the variables as well as you wished, how could you do it better next time? What could you do that would further your research into this topic? If you found out that gray water was okay for your houseplants, how could you expand the project? Could you collect enough gray water to water a garden?

The final part of your report is the Bibliography. You will be very glad you kept track of all your sources of information. Here is where you list them. Ask your teacher just how they should be presented.

THE DISPLAY

For many students, making the display is a lot of fun. You get a chance to be creative and colorful. Check with your teacher about the size of your display. International Science and Engineering Fair rules say your display should stand by itself. The largest it can be is 48 inches (122 cm) wide, 30 inches (76 cm) deep, and 108 inches (274 cm) high. If your display will sit on a table, you must count in about 30 inches (76 cm) for the table height. Your school fair may have different rules. Check with your teacher. And remember that you have to

transport the display. Don't make it bigger than your car can hold.

Your *backdrop* will hold your display. If you're in elementary school, you can cut off the top and one side of a large cardboard box for your backdrop. If you're older, you will want to make it better looking.

One of the easiest and lightest backdrops can be made from Foamkore sheets taped together with duct or book tape. We used a backdrop made from wood and Masonite board several years before Foamkore was available. It lasted a long time and was very sturdy, but it was also very heavy.

On your backdrop you will present your work in the most colorful, interesting way you can. More and more, science fairs are asking that students not bring their materials for display. Many things you might have used are not allowed at all in a display. Check with your teacher.

It is probably best if you don't display your gerbil and the maze he ran. Instead, you can take photos of the experimental runs. Add your photos as they apply to each section of the display.

The first thing visitors and judges will see is your title. It needs to be easily read and interesting. It is also important that it reflect what you actually tested. You will probably put your title on the middle of the back section of your backdrop.

On the left section of the backdrop, you will put the Purpose and below that, your Hypothesis. Make these statements simple, but complete.

Below your title, you will outline your Procedure. On the right side, put your Results (more on this later) and your Conclusion. The experiment report and your laboratory notebook can be displayed in the open space created by the three sides.

Charts or graphs that you can prepare to display your data (Results) will help make your project clear to others. These can be displayed under the Results section along with a brief summary of the Results. Use your photographs to illustrate what you did. You can add them to the backdrop or have them in an album on the tabletop.

By adding color to your graphs or using colored poster board or paper behind each section, you can make your project more appealing. You may want to use fabric to cover the tabletop. Use your imagination and good taste. It is most important that it be easy to read. You can have it printed on a computer or print it yourself if you write well. Make lines to guide you if you do it yourself.

Remember, color and computer printouts may make your display appealing, but they won't help a project that's not done well. On the other hand, a sloppy presentation will hurt a good project.

THE JUDGING

Judges are most interested in finding out if you understand what you did. If your father is a scientist and did the work of your project for you, you will have a wonderful project and display, but you won't be able to convince the judge that you did it. Your adult partner is there to help you, not to do the work for you.

The judge will probably ask you to tell about your project. Practice this at home. Tell an adult about your project. Begin with why you wanted to study what you did. Then mention your hypothesis and how you tested it. Be sure you mention your results and what they mean.

There are some things you will need to remember when you talk with judges. Be polite. The judge is there because he or she is interested in science and students. They do not get paid for judging. Thank the judge for the time he or she spent with you before the judge moves on.

Don't chew gum while talking to the judge. Speak clearly and look at the judge. Refer to your display and your photographs, but don't just read the printed material. Smile and let the judge know you enjoyed doing your project.

10.
The Projects

The projects in this book will get you started. If you've never done a science fair project before, you may want to follow one here. You may want to create your own. Creativity is important when you get to junior high and high school science fairs. Make yours unique, even if you use one listed here as a starting point.

Your research will lead you to new ideas. Ask yourself, "What if I make this change in the project?" Scientists are creative people, looking for ways to make life better, or for ways to understand life better. Perhaps you will find out something new.

IS GRAY WATER GOOD FOR MY PLANTS?
Botany/Environmental Science

PURPOSE: to determine if using gray water is beneficial to houseplants

MATERIALS: four houseplants such as *pothos*, the same size and age; four identical pots; potting soil; measuring cup; gray water; tap water

PROCEDURE:
1. Develop your hypothesis from your research.
2. Plant the four plants and put in a place that is appropriate for that type of plant. Make sure each plant has a saucer under it. Mark two plants *A* and two plants *B*.
3. Water plants *A* with tap water as directed or needed. Water plants *B* with gray water in the same amounts as plants *A*. Measure the amount given each plant. Keep everything the same.
4. Count the number of new leaves the plant has put on.
5. Analyze your data.
6. Accept or reject your hypothesis and tell why.

REMEMBER: Keep all variables except the type of water the same for all four plants.

KEY WORDS: gray water, arid gardening, plant name

THINGS TO THINK ABOUT: Could you design a house that uses its gray water effectively? What regions of our country and the world might benefit from gray water irrigation?

WHAT SNACKS SOLD AT
THE SCHOOL SNACK BAR ARE
LOWEST IN EMPTY CALORIES?
Medicine and Health

PURPOSE: to determine which snacks sold at my school are lowest in empty calories

MATERIALS: various snacks from the snack bar

PROCEDURE:
1. Develop your hypothesis from your research.
2. Using your knowledge of nutrition and the content of fat, sugar, vitamins, minerals, and fiber of foods, list the nutritional components of each snack.
3. Decide which snack is lowest in empty calories.

REMEMBER: A snack should provide only a portion of nutrients a person needs per day. If a snack is too heavy on empty calories, it will be unhealthy.

KEY WORDS: nutrition, nutritional information, snacks, fat, empty calories, vitamins, sugar, specific food items

THINGS TO THINK ABOUT: What happens if students don't eat balanced meals? What can happen in later life if a student develops poor eating habits? How do the nutritional needs of a student your age and those of an adult vary? Americans eat too much fat and one of three adults in the United States is overweight. Is this a cause-and-effect relationship? What can be done to help the general population eat better? Can you do the above experiment and decide which snacks are highest in empty calories? Why does your snack bar sell them?

CAN I BUILD A BETTER
HUMMINGBIRD FEEDER?
Zoology

PURPOSE: to determine the best shape, size, and color of a hummingbird feeder

MATERIALS: will vary depending on your design

PROCEDURE:
1. Develop your hypothesis from your research.
2. Using your knowledge of hummingbirds and hummingbird behavior, design a hummingbird feeder.
3. Make a model if you can. Make drawings if you can't make the model.

REMEMBER: You will have to study hummingbird behavior around a typical hummingbird feeder. Some things you will need to think about include the aggressive rufous hummingbird, ants, bees, and wasps who are interested in the nectar, and how many hummingbirds you have visiting your feeder at once.

KEY WORDS: hummingbird, rufous, hummingbird feeders

THINGS TO THINK ABOUT: Can you design a yard plan that takes into account the aggressive behavior of some hummingbirds? Which nectar solution will the birds prefer?

DO SENIOR CITIZENS EAT BALANCED MEALS?

Medicine and Health/Behavioral Science

PURPOSE: to determine if senior citizens I know eat balanced diets

MATERIALS: survey

PROCEDURE:
1. Develop your hypothesis from your research.
2. Develop a survey that will ask senior citizens what they eat during a typical day.
3. Analyze the data.
4. Accept or reject your hypothesis.

REMEMBER: Your survey needs to be specific. It also needs to be simple. Can you make a check sheet for the senior citizens to fill out? Do you want to give the survey in person?

KEY WORDS: elderly nutrition, senior citizens and health issues, nutritional needs

THINGS TO THINK ABOUT: What patterns show up in senior citizens' eating habits? What happens to the bodies and lifestyles of people as they age? What do these things mean to nutritionists? The population of the United States is aging. What can be done for future senior citizens? You might consider interviewing someone on the staff at a nursing home to get other ideas on problems of the elderly.

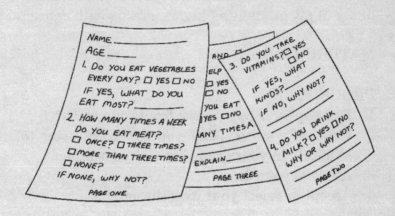

NAME _____
AGE _____
1. Do you eat vegetables every day? ☐ YES ☐ NO
   IF YES, WHAT DO YOU EAT MOST? _____
2. How many times a week do you eat meat?
   ☐ ONCE? ☐ THREE TIMES?
   ☐ MORE THAN THREE TIMES?
   ☐ NONE?
IF NONE, WHY NOT?
PAGE ONE

AND ☐
ELP
☐ YES
☐ NO
you eat
☐ YES ☐ NO
MANY TIMES A
EXPLAIN _____
PAGE THREE

3. DO YOU TAKE VITAMINS? ☐ YES ☐ NO
   IF YES, WHAT KINDS? _____
   IF NO, WHY NOT? _____
4. DO YOU DRINK MILK? ☐ YES ☐ NO
   WHY OR WHY NOT? _____
PAGE TWO

DO STUDENTS AT MY SCHOOL KNOW SOMEONE WITH CANCER?
Medicine and Health

PURPOSE: to determine the extent to which students in my grade level know persons with cancer

MATERIALS: surveys

PROCEDURE:
1. Develop your hypothesis from your research.
2. Develop a survey that you can administer to students in your grade level at your school.
3. Analyze your data.
4. Accept or reject your hypothesis.

REMEMBER: You need to get a sample of all the students in your grade at your school. Remember that the larger the sample, the greater the accuracy of your results.

KEY WORDS: cancer, disease awareness

THINGS TO THINK ABOUT: Could you do the same thing for persons with AIDS/HIV? What happens to people who are infirm or diseased? Can your friends tell if someone has cancer? Can they tell if someone has AIDS?

DO STUDENTS AT MY SCHOOL
EAT BREAKFAST?
Behavioral Science

PURPOSE: to determine if students in my grade level at my school eat breakfast

MATERIALS: surveys

PROCEDURE:
1. Develop your hypothesis from your research.
2. Develop a survey to administer to students in your grade level at your school.
3. Decide what your sample will be.
4. Give the survey and collect data.
5. Analyze your results.
6. Accept or reject your hypothesis.

REMEMBER: You will want to get a sampling of students, not just one class, unless you focus your study on just your class.

KEY WORDS: breakfast, nutrition, learning

THINGS TO THINK ABOUT: It will be important to tie your study into nutritional needs and how well students do in school. You could also survey students to see if they understand

why breakfast is important. Can you think of
other surveys?

ARE THE STUDENTS AT MY SCHOOL AWARE OF SMOKING FACTS?
Medicine and Health

PURPOSE: to determine if students at my school are aware of smoking facts and dangers

MATERIALS: surveys

PROCEDURE:
1. Develop your hypothesis from your research.
2. Develop a survey to administer to students at your school.
3. Decide how you will get a sample.
4. Give the survey and collect data.
5. Analyze your results.
6. Accept or reject your hypothesis.

REMEMBER: The more students who answer your survey, the more likely your results will be accurate.

KEY WORDS: smoking, effects of smoking, teenagers and smoking

THINGS TO THINK ABOUT: You might decide to survey students to see if they know facts about AIDS or drugs. What else do the students at your school need to know? You could also survey the students to see if any of them smoke. Or survey adults who smoke and find out how old they were when they started smoking.

WHAT IS THE AVERAGE FAMILY SIZE
OF STUDENTS AT MY SCHOOL?
Behavioral Science

PURPOSE: to determine the average family size for students in my school

MATERIALS: surveys

PROCEDURE:
1. Develop your hypothesis from your research.
2. Develop a survey to administer to students in your school.
3. Decide how you will get a sample.
4. Give the survey and collect the data.
5. Analyze your results.
6. Accept or reject your hypothesis.

REMEMBER: A larger sample will give you a more accurate picture of students' family size for your school. Could you obtain this information from the administration?

KEY WORDS: family size, census, census taking, Zero Population Growth (ZPG)

THINGS TO THINK ABOUT: The parents of children your age are probably baby boomers.

During the late 1960s, a group was formed advocating Zero Population Growth (ZPG). Is this reflected in the family size of students at your school? What other things might be factors?

HOW COMMON ARE HEADACHES
IN STUDENTS AT MY SCHOOL?
Medicine and Health

PURPOSE: to determine what percentage of students in my grade level at my school get headaches

MATERIALS: surveys

PROCEDURE:
1. Develop your hypothesis from your research.
2. Develop a survey to administer to students in your grade level at your school.
3. Decide how you will get a sample.
4. Give the survey and collect the data.
5. Analyze your results.
6. Accept or reject your hypothesis.

REMEMBER: Keep in mind that you cannot get a response from every student in your school, unless your school is very small.

KEY WORDS: headaches, migraines, proliferation of headaches, headaches in children

THINGS TO THINK ABOUT: Why do people get headaches? Why do some people get terrible headaches and others not get headaches at all? What are some of the reasons students in your school get headaches?

WHAT KINDS OF PETS DO THE
STUDENTS AT MY SCHOOL HAVE?
Zoology

PURPOSE: to determine which is the most popular type of pet with students in my grade level at my school

MATERIALS: surveys

PROCEDURE:
1. Develop your hypothesis from your research.
2. Develop a survey to administer to students in your school.
3. Decide how you will choose your sample.
4. Give the survey and collect the data.
5. Analyze your results.
6. Accept or reject your hypothesis.

REMEMBER: You will have to ask very specific questions. If you want to know what type of pet your classmates would like to have, word your survey to get data on preferences. If you want to know what types of pets your classmates have, ask questions that will result in data about pets they already own. Are reptiles popular with students at your school?

KEY WORDS: pets, dogs, cats, Humane Society, pet populations, reptiles as pets

THINGS TO THINK ABOUT: Many animals are destroyed every year because too many cats and dogs are born. Do most of the dog owners have purebred or mixed breed dogs? Did they get them from a pet store or a breeder or the animal shelter or Humane Society? What can be done about over-population of pets? Do the students in your school have pets who are spayed or neutered? There are many different ways to approach this problem.

WHAT IS THE EFFECT OF HEAT ON ALGAE GROWTH?
Environmental Science

PURPOSE: to determine the effect of heat on the growth of algae in a fish tank

MATERIALS: two fish tanks or bowls, pond or fish tank water, two fish tank heaters, two fish tank thermometers

PROCEDURE:
1. Develop your hypothesis from your research.
2. Controlling all the variables except temperature, fill two tanks or bowls with pond or fish tank water.
3. Raise the temperature of one tank.
4. Allow the tanks to sit at a constant temperature. Note changes in your project notebook.
5. When does algae begin to form on the sides of the glass?
6. At the end of the experiment you will have to decide how you will measure the difference in algae growth.
7. Accept or reject your hypothesis.

REMEMBER: You will need some way to measure the amounts of algae. How can you do this?

KEY WORDS: algae, global warming, changes in sea temperature, plankton and algae, algae and oxygen levels

THINGS TO THINK ABOUT: What might be a change in the oceans if the earth warms and glaciers melt? Will the rising temperatures change the amount of algae now in the oceans? What will this mean for plankton, fish, and other ocean dwellers? What happens in a fish tank or pond when the algae growth is very heavy?

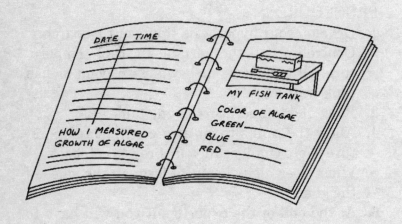

WHAT IS THE EFFECT OF DECREASED SALINITY ON BRINE SHRIMP?
Zoology

PURPOSE: to determine if a decrease in salinity has an effect on the numbers in brine shrimp populations

MATERIALS: two tanks or bowls, brine shrimp, saline water

PROCEDURE:
1. Develop your hypothesis from your research.
2. Add 5 percent tap water to one tank of brine shrimp and 5 percent saline water to the other. Mark the tanks.
3. After a week, two weeks, and a month, take a sample and count the brine shrimp.
4. Record your results in your project notebook.
5. Accept or reject your hypothesis.

REMEMBER: When you add anything to the tank, it must be the same temperature as the tank water to control the variable of heat. Make sure you learn how to care for brine shrimp.

KEY WORDS: brine shrimp, salinity of water, diluting the oceans, global warming and melting polar caps, aquariums

THINGS TO THINK ABOUT: You could also test the variable of temperature. Why are brine shrimp important?

WHAT IS THE EFFECT
OF STORMS ON POLLUTION?
Environmental Science

PURPOSE: to determine if storms have an effect on air quality

MATERIALS: air quality data, weather data

PROCEDURE:
1. Develop your hypothesis from your research.
2. In your notebook, keep track of the weather data and the air quality data.
3. Make a chart showing the relationship of pollutants in the air with the weather changes.
4. Measure your results.
5. Accept or reject your hypothesis.

REMEMBER: Accuracy in keeping track of the information on a daily basis is important.

KEY WORDS: storms, weather changes, wind velocity, pollution, air quality, rain and pollution, air pressure

THINGS TO THINK ABOUT: What can or should be done to educate people about the relationship between storms and pollution?

DOES THE WIND AFFECT
CARBON MONOXIDE LEVELS?
Environmental Science

PURPOSE: to determine if wind conditions affect levels of carbon monoxide in my town

MATERIALS: access to wind, weather, and carbon monoxide levels

PROCEDURE:
1. Develop your hypothesis from your research.
2. Chart the wind conditions and the carbon monoxide levels for your area for a month.
3. Graph your results.
4. Accept or reject your hypothesis.

REMEMBER: Keeping accurate records will be very important for this type of study. It is a study that will work best if you live in a city. However, if you can get the data from a city's newspaper or off the Internet, you can do this project from anywhere.

KEY WORDS: carbon monoxide, wind, weather, greenhouse effect

THINGS TO THINK ABOUT: Can you measure the carbon dioxide/carbon monoxide levels in the air inside your home? Can you obtain information from the companies that make detectors?

DOES FALSE NIGHT AFFECT THE BIRTHRATE IN BATS?
Zoology

PURPOSE: to determine if false night affects the birthrate in bats

MATERIALS: bat birthrate data

PROCEDURE:
1. Develop your hypothesis from your research.
2. Visit a zoo with a display of nocturnal animals under a red light to simulate night. Find out what the zoo officials do to simulate day during the night.
3. Research the birthrate of one species, such as a type of bat. Find out if the false night affects the birthrate.
4. Accept or reject your hypothesis.

REMEMBER: This is a research project. You will not be raising bats in false night conditions. Your research must be very thorough. Talk with zoo officials and with animal experts at your state university.

KEY WORDS: zoos, nocturnal animals, bats

THINGS TO THINK ABOUT: Zoos are changing the cycle of nocturnal animals so visitors can view the animals in their active time. Is this a good idea for the animals?

DOES SHAVING BODY HAIR
REDUCE DRAG ON SWIMMERS?
Physics

PURPOSE: to determine if shaving body hair reduces drag on, and consequently adds speed to, competitive swimmers

MATERIALS: stopwatch or official timing mechanism, pool, swimmers

PROCEDURE:
1. Develop your hypothesis from your research.
2. Using several swimmers, time them swimming their favorite race event.
3. Ask the swimmers to shave their body hair.
4. The next day, or later the same day, but not too soon after their first swim, time them again.
5. Accept or reject your hypothesis.

REMEMBER: You will need a way to control this experiment as much as you can. Talk with a swimming coach about how to do this.

KEY WORDS: swimming, drag, body hair

THINGS TO THINK ABOUT: Swimmers are stronger and faster than ever before. Why?

DO OVER-THE-COUNTER MEDICATIONS HELP THE COMMON COLD?
Medicine and Health

PURPOSE: to determine if taking over-the-counter cold remedies helps to cure a cold

MATERIALS: access to research

PROCEDURE:
1. Develop your hypothesis from your research.
2. Interview your family doctor as well as getting information from other sources.
3. Talk with people who claim to have several colds a year. What do they think?
4. Accept or reject your hypothesis.

REMEMBER: This is a research project. You will have to be careful and thorough in your research.

KEY WORDS: cold remedies, common cold

THINGS TO THINK ABOUT: It will be interesting to talk with people to find out what they think. It would also be interesting to analyze television and magazine ads for cold remedies. These will tell you what people think and are told, but they will not necessarily be factual. Compare this information with the facts you obtain from your research. Colds strike most Americans every year. Can you find out how many workdays are missed due to the common cold? What do the experts recommend to avoid colds? What can you and other students do about colds? Is having a "cold" and having a "virus" the same thing?

DO STORMS AFFECT
MIGRAINE SUFFERERS?
Medicine and Health

PURPOSE: to determine if weather affects migraine headaches

MATERIALS: subjects who suffer from migraines, calendars, access to weather information

PROCEDURE:
1. Develop your hypothesis from your research.
2. Find several migraine sufferers. From your research, you will know who are the most likely people. Ask your family doctor to ask his migraine-suffering patients if they would participate in the study.
3. Give the subjects a calendar on which to mark days they suffered from migraines. Have them use two colors of pen, one for mild and one for severe migraines.
4. Collect weather data for the same period.
5. Combine the health data and compare with the weather data.
6. Accept or reject your hypothesis.

REMEMBER: You should have about ten people in your study. Most medical studies use hundreds of subjects and follow them for a year or so. Obviously you can't do anything on that scale. But you might find out something interesting. Let your subjects know what your findings are.

KEY WORDS: migraines, weather patterns, storm fronts

THINGS TO THINK ABOUT: Don't let the subjects know you are watching weather while they are recording their headaches. After the study, tell them what you were studying and what you found out. Do you know why? People's minds are highly susceptible to suggestion. You might be triggering headaches for them just because the weather is changing. Migraines are incapacitating to those who get them.

Could you do the same experiment for tension headaches? What about blood sugar levels and headaches?

CAN I DESIGN A SOLAR HOME?
Engineering

PURPOSE: to determine what materials and configurations would best take advantage of solar energy

MATERIALS: depends upon what you find out from your research — you may just draw the design, or you may prefer to make a model or a model of a special wall

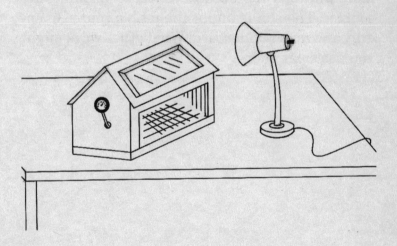

PROCEDURE:
1. Design a home that incorporates the solar ideas you have found in your research.
2. Decide how you will present your design.

REMEMBER: You will need to do thorough research in order to understand solar design.

KEY WORDS: solar design, solar homes, home plans

THINGS TO THINK ABOUT: We are becoming more aware of the need to conserve our finite resources. The sun is an inexhaustible source of energy and much underused. What other ways might you use solar energy?

CAN I DESIGN A BETTER
BICYCLE HELMET?
Engineering

PURPOSE: to determine what makes a good bicycle helmet and design one that takes these things into consideration

MATERIALS: will depend on what you decide to use in making your helmet

PROCEDURE:
1. Research bicycle helmet design.
2. Design a bicycle helmet that takes into consideration comfort, safety, and features that you would like.
3. Decide how you will present your design, either in drawings or in a model.

REMEMBER: Bicycle helmets must be comfortable as well as safe or people will not wear them. What can you do to yours to make it comfortable? Consider weight in your design.

KEY WORDS: bicycle helmets, head injuries

THINGS TO THINK ABOUT: Bicyclists are sometimes hit by cars they never see. Can you do something about this in your design?

IS THE AVERAGE TEMPERATURE
OF MY CITY WARMING?
Environmental Science

PURPOSE: to determine the trend of average temperatures in my city

MATERIALS: access to information about temperatures

PROCEDURE:
1. Develop your hypothesis from your research.
2. Chart average temperatures over a period of time. Compare them with average temperatures of the same period ten, twenty-five, and fifty years earlier.
3. Compare your results.
4. Accept or reject your hypothesis.

REMEMBER: The more data you can amass, the more accurate your picture of average temperatures will be.

KEY WORDS: global warming, temperatures, (city name)

THINGS TO THINK ABOUT: What would rising temperatures mean to your area? How would

they change the agriculture near you? Who would
suffer most?

CAN I KEEP THE BEACH
FROM WASHING AWAY?
Environmental Science

PURPOSE: to determine what will keep the beach from washing away with the action of the water and weather

MATERIALS: sand, electric fan, water, something to simulate waves, something to prevent waves from washing the beach

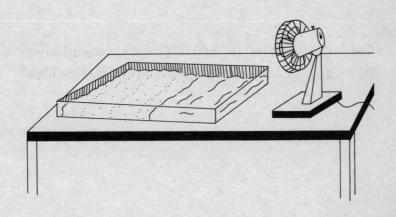

PROCEDURE:
1. Develop your hypothesis from your research.
2. Make a model of beach and waves and wind.
   Decide what you will add to stop the washing
   action.
3. Test your idea.
4. Accept or reject your hypothesis.

REMEMBER: The more realistic you can make
your model, the more your answer will be applica-
ble.

KEY WORDS: erosion, beach erosion, wave action

THINGS TO THINK ABOUT: Who would bene-
fit from your idea? Is it a financially feasible idea?

CAN I SOUNDPROOF MY ROOM?
Engineering

PURPOSE: to determine how to soundproof a bedroom in an existing structure

MATERIALS: will depend on your ideas and your display

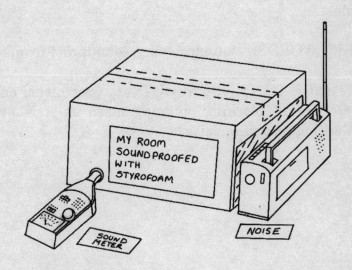

PROCEDURE:
1. Develop your hypothesis from your research.
2. Make a model of your room. Try soundproofing it with materials you think will work.
3. Test your model.
4. Accept or reject your hypothesis.

REMEMBER: Test your model both before and after you have added soundproofing to see how well your idea works. Use a sound meter if possible.

KEY WORDS: soundproofing, soundproof rooms

THINGS TO THINK ABOUT: Is your model as effective at blocking outgoing noise as it is at blocking incoming noise?

CAN I DESIGN AN AERODYNAMIC CAR?
Physics

PURPOSE: to determine if a car body that is better designed will encounter less wind resistance

MATERIALS: wind tunnel, material to model car

PROCEDURE:
1. Develop your hypothesis from your research.
2. Design and build a car that is as aerodynamic as you can make it.

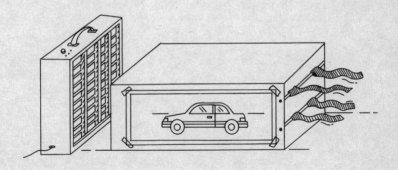

3. Build a wind tunnel large enough for your car.
4. Test your car. Test another car of similar size, but not aerodynamically designed.
5. Photograph your experiments.
6. Accept or reject your hypothesis.

REMEMBER: You may need to add smoke to your wind tunnel to make the airstream visible.

KEY WORDS: wind tunnel, cars and aerodynamics, fuel efficiency

THINGS TO THINK ABOUT: Possibilities for variations of this project include designing a solar-powered car. Could you design an aerodynamic motor home?

HOW MANY HOURS A DAY
DO PEOPLE WATCH TV?
Behavioral Science

PURPOSE: to determine the average number of hours per day students from my school and their parents watch television

MATERIALS: calendars to give to students and their parents to keep track of their television viewing

PROCEDURE:
1. Develop your hypothesis from your research.
2. Find sufficient subjects willing to write down how many hours a day they watch television over a period of time.
3. Collect your data. Analyze your results.
4. Accept or reject your hypothesis.

REMEMBER: Try to get a large number of the students in your classroom to participate in this study. Compare the hours a day the student watches TV with the hours a day a parent watches TV.

KEY WORDS: television viewing, television and children

THINGS TO THINK ABOUT: SAT scores for high school students wanting to go to college are dropping from levels of twenty-five years ago. Why? You could also find out how many hours your friends play computer or video games, how much time they spend on the Internet, how much time they spend reading for pleasure, and how much time they spend on homework.

HOW CAN WE SAVE THE HONEYBEES?
Zoology/Environmental Science

PURPOSE: to determine why some bee colonies are dying out and what to do about it

MATERIALS: access to research on honeybees

PROCEDURE:
1. This is a research project. First you will want to find out about honeybees. Next, you will want to know what is happening to some honeybee colonies.
2. When you understand the relationship of bees to the cause of their destruction, try to imagine what could be done to stop the deaths.

REMEMBER: Find out first all about honeybees. They are important beyond giving us honey.

KEY WORDS: honeybees, honeybee colonies, killer bees, mites

THINGS TO THINK ABOUT: How would our lives be different without honeybees? What could replace them?

CAN I DESIGN A BETTER FOOTBALL?
Physics

PURPOSE: to determine if another shape or material would be more accurate in tests that involve throwing the ball ten yards to a target

MATERIALS: will depend on your design

PROCEDURE:
1. Develop your hypothesis from your research.
2. Try making balls of different shapes from materials such as papier-mâché, foam, Styrofoam, etc.
3. Have three friends help you test the balls by throwing them at a target ten yards away. Record the accuracy of their attempts.
4. Accept or reject your hypothesis.

REMEMBER: Football players need to be able to throw, kick, and carry the ball. A handle might make a football easier to carry, but it would make it harder to throw accurately and to kick.

KEY WORDS: aerodynamics, balls, footballs

THINGS TO THINK ABOUT: How have footballs changed over the years? Could you design a

ball and invent a game to go with it? Could you
use a soccer ball for playing football?

CAN I JUMP HIGHER IN THE LATEST ATHLETIC SHOES?
Physics

PURPOSE: to determine if new designs in athletic shoes will increase jumping capacity

MATERIALS: ordinary sneakers, newly designed athletic shoes

PROCEDURE:
1. Develop your hypothesis from your research.
2. Measure how high you and your friends can jump with your ordinary sneakers or school shoes.
3. Put on newly designed athletic shoes. Measure how high you can jump now.
4. Collect your data. Analyze your results.
5. Accept or reject your hypothesis.

REMEMBER: Give each person several trials in each pair of shoes. Have them put a hand in acrylic paint of one color for one type of shoe and then a different color for the other type of shoe. Have them jump and touch butcher paper you have attached to a wall. Measure all the handprints. If the bottom of your paper is level with the floor, you can measure the paper after you take it

down. Measure to the same part of every hand-
print.

KEY WORDS: high jumping, athletic shoes

THINGS TO THINK ABOUT: Can you design a
shoe that would help you run faster or jump
higher? How would it be different from what you
now have?

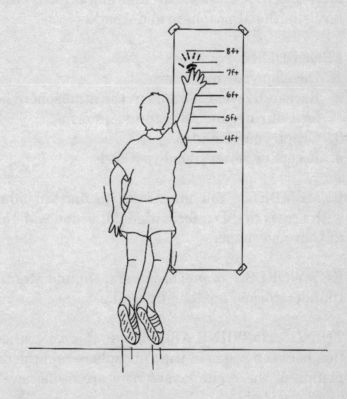

DO THUNDERSTORMS
AFFECT OZONE LEVELS?
Environmental Science

PURPOSE: to determine if levels of ozone in the atmosphere increase or decrease in relation to thunderstorms

MATERIALS: access to information on ozone levels in the atmosphere and weather

PROCEDURE:
1. Develop your hypothesis.
2. Research levels of ozone in the atmosphere before, during, and after thunderstorms.
3. Graph your findings.
4. Accept or reject your hypothesis.

REMEMBER: You may want to find an adult with access to this information if it is not available in your newspaper.

KEY WORDS: ozone, ozone and thunderstorms, thunderstorms, weather, lightning

THINGS TO THINK ABOUT: Is there a connection between ozone in the atmosphere and the depletion of the ozone layer? Why are ozone levels monitored?

CAN I MAKE A ROBOT
TO CLEAN MY ROOM?
Physics/Engineering

PURPOSE: to put together a remote control vacuum cleaner

MATERIALS: remote control, chassis and wheels, small vacuum cleaner, other parts as necessary

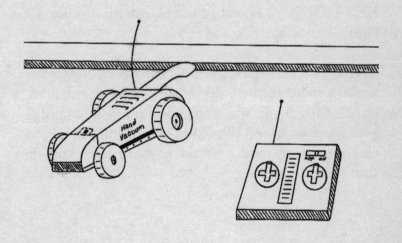

PROCEDURE:
1. Research how a remote control car is put together.
2. Design and build a remote control vacuum.
3. Try it out.
4. Take plenty of photographs to display.

REMEMBER: You need to understand how the remote control works. Take into account the weight of the small vacuum in your design.

KEY WORDS: remote control, robotics, robot design

THINGS TO THINK ABOUT: Could you ever make a robot that would be able to distinguish between things you want to throw away and things you want to keep? What takes you longest about cleaning your room? Could a robot help you do that?

DOES DIET SODA AFFECT MY TEETH?
Medicine and Health

PURPOSE: to determine if sugarless sodas will affect human teeth

MATERIALS: sodas, jars, teeth

PROCEDURE:
1. Develop your hypothesis from your research.
2. Set up four jars. Add a tooth to each. Maybe your parents kept your baby teeth. If not, call your dentist for suggestions.
3. In one jar, add 12 oz tap water. In the second jar, add a 12 oz sugarless soda. In the third jar, add a 12 oz soda that is not sugarless. The fourth jar should contain 12 oz of carbonated water.
4. Label the jars and check them, making notes in your project notebook. Gather the data and analyze it.
5. Accept or reject your hypothesis.

REMEMBER: Your jar of plain water is your control. Does plain water affect the tooth?

KEY WORDS: tooth decay, carbonated beverages and teeth, sugar and tooth decay

THINGS TO THINK ABOUT: How much soda do you drink in a day or week? Do you drink regular or diet sodas? Why? Which would your dentist recommend?

WHEN DO PEOPLE HAVE THEIR WISDOM TEETH PULLED?
Medicine and Health

PURPOSE: to determine the age at which adults have their wisdom teeth pulled

MATERIALS: surveys

PROCEDURE:
1. Develop your hypothesis from your research.
2. Design a survey to administer to a group of adults.
3. Collect and analyze your results.
4. Accept or reject your hypothesis.

REMEMBER: You will need to research statistics on wisdom teeth. A medical library would be a great deal of help. If you don't have one near you, you can talk with your dentist. Your dentist probably has journals with research on wisdom teeth.

KEY WORDS: wisdom teeth, oral surgery

THINGS TO THINK ABOUT: Why do you think people have wisdom teeth pulled? Why are they called wisdom teeth? You might ask if you can survey the teachers in your school.

DOES EGGSHELL WATER AFFECT
GROWTH OF HOUSEPLANTS?
Botany

PURPOSE: to determine if water that has held empty eggshells has an effect on the growth of houseplants

MATERIALS: large jar; tap water; eggshells; two identical coleus plants, potted, with saucers under them

PROCEDURE:
1. Develop your hypothesis from your research.
2. Set the two coleus plants in a good location for light, heat, and to be safe from interference.
3. Controlling the variables, water one plant with tap water and the other with tap water in which you have soaked eggshells from raw eggs.
4. Chart the growth and note the differences in your notebook.
5. Accept or reject your hypothesis.

REMEMBER: Decide before you begin how you will measure the differences in the plants. Will it be height? Number of leaves? Be sure you label the pots: TAP WATER, EGGSHELL WATER.

KEY WORDS: eggshells, albumen, plant growth, coleus, calcium and plants

THINGS TO THINK ABOUT: Could you use a different plant than coleus? What about evergreens? Would eggshells from hard-boiled eggs be the same?

WHAT IS THE EFFECT OF ROOT STIMULATOR ON GERANIUM CUTTINGS?
Botany

PURPOSE: to determine the effect of root stimulator on the development of roots of geranium cuttings

MATERIALS: clean jars, cuttings of equal size from a geranium, root stimulator

PROCEDURE:
1. Develop your hypothesis from your research.
2. Add the root stimulator to a quart of water in the proportion suggested on the label.
3. Put each cutting in a jar and label the jars.
4. Add tap water to one jar so the end of the cutting of the geranium is in about an inch of water.
5. Add the same amount of water with root stimulator to the second jar.
6. Decide how you will measure the differences. Collect data in your notebook. Analyze the data.
7. Accept or reject your hypothesis.

REMEMBER: Keep all the variables the same except the type of water added to the jars.

KEY WORDS: root stimulator, cuttings, geranium, propagation by cuttings

THINGS TO THINK ABOUT: You could use cuttings from other types of plants. What other types of plants can propagate by cuttings? Talk with a nursery person at the garden center. Do they use root stimulator on their new plantings? Do any plants not respond to root stimulator?

IS HOT WATER BETTER THAN COLD FOR WASHING DISHES?
Microbiology

PURPOSE: to determine if hot water reduces bacteria on plates better than cold water

MATERIALS: plates, agar in petri dishes, distilled water, eye dropper, hot and cold water

PROCEDURE:
1. Develop your hypothesis from your research.
2. Pass four plates around to your family. Have them touch the plates all over the food surface.
3. Wash two plates as you would normally with hot water and two with cold water. Measure the temperature of the water.
4. Boil and cool a quart of distilled water. Add 2 cc distilled water to each plate. Let it run across the plate and into an agar-filled petri dish. Cover and label the petri dishes: 2 HOT and 2 COLD.
5. Add 2 cc distilled water to the fifth agar-filled petri dish as a control. Cover and label it, CONTROL.
6. Compare the amounts of bacteria that grow in the petri dishes.
7. Analyze your results and accept or reject your hypothesis.

REMEMBER: Be careful not to contaminate the petri dishes. Make sure you know how to dispose of the used petri dishes in a safe manner.

KEY WORDS: heat and bacteria, hot water and bacteria

THINGS TO THINK ABOUT: There are many things you can test for bacteria in this way. You could ask what is the dirtiest surface in your house. How would you structure a project to test that?

CAN YOU TEACH AN
OLD DOG NEW TRICKS?
Zoology

PURPOSE: to determine if an old dog can learn a new trick

MATERIALS: old dog, treats, leash

PROCEDURE:
1. Develop your hypothesis from your research.
2. Design a training session for your old dog.
3. Take photographs. Ask the dog to perform the trick before you have trained the dog.
4. Train the dog using your newly designed technique.
5. Ask the dog to perform the trick again. Take a photograph.
6. Keep notes on how many sessions it took to teach your old dog a new trick. Or, notes on how many sessions you tried before you gave up.
7. Accept or reject your hypothesis.

REMEMBER: Before you work with animals, you need permission from your teacher. If you are in middle school or junior high school, you may need permission from the regional science office. Your teacher can get a form for you.

KEY WORDS: dog training, obedience classes

THINGS TO THINK ABOUT: Why do people say you can't teach an old dog new tricks? Can you train a cat? Can you train your pet reptile? How would training for a cat or a reptile be different from training for a dog? Would a reptile or a cat respond to words of praise?

IS SEEING BELIEVING?
Computers

PURPOSE: to determine if people can detect changes made by computers to their family photographs

MATERIALS: computer, graphics program, scanner

PROCEDURE:
1. Develop your hypothesis from your research.
2. Scan a family photograph into your computer. Using a morphing/graphics program, make a small change in one person, such as the color of the eyes or hair.
3. Print both photographs with a color printer and mount them.
4. Ask a group of subjects to see if the photographs are the same.

REMEMBER: You will probably want to work with your computer teacher or a computer programmer on this project.

KEY WORDS: computer imaging, morphing

THINGS TO THINK ABOUT: How will you present your findings at a science fair? Does your school have a computer that you can use to run your program and display it?

CAN I GROW PLANTS
IN MARTIAN SOIL?
Astronomy/Botany

PURPOSE: to determine if plants will grow in nonorganic soil

MATERIALS: potting soil, seeds (beans, peas, corn, or marigolds are good choices), pots

PROCEDURE:
1. Develop your hypothesis from your research.
2. Destroy the organic content in half the potting soil by baking it in a self-cleaning oven during the cleaning cycle. Let it cool.
3. Plant four seeds, two in the baked soil and two in regular potting soil. Label the pots.
4. Keep all the variables the same. Water and care for the seeds.
5. Keep notes on differences in germinating time in your notebook.
6. Decide how you will determine the difference you want to measure. Gather your results.
7. Accept or reject your hypothesis.

REMEMBER: It will be important to control the variables carefully in this experiment. Only the soil should be different.

KEY WORDS: germination, soil quality, soil ingredients, bacteria and plants, Mars and soil

THINGS TO THINK ABOUT: How could you make the baked soil normal again? What would you add?

You might use a soil with a greater sand or clay content than potting soil and have your teacher bake it in a ceramics kiln.

Check out the NASA sites on the Internet. See if you can find the latest research from NASA.

About the Author

Penny Raife Durant is the author of six science books for children, including the predecessor to this book, *Prizewinning Science Fair Projects,* and the award-winning novel for young adults, *When Heroes Die.* She is a former elementary school teacher and preschool teacher and director. She has B.S. and M.A. degrees in education from the University of New Mexico. She lives with her husband and her son in Albuquerque.

Paul Mitschler, consultant to the book, has been a teacher of science in the Topeka and Albuquerque schools for thirty-two years, spanning all grade levels from kindergarten through high school. He is currently a teacher of the gifted at Jefferson Middle School in Albuquerque. He has B.S. and M.A. degrees from the University of Kansas.